ICME-13 Topical Surveys

Series editor

Gabriele Kaiser, Faculty of Education, University of Hamburg, Hamburg, Germany

More information about this series at http://www.springer.com/series/14352

Gerald A. Goldin · Markku S. Hannula ·
Einat Heyd-Metzuyanim · Amanda Jansen ·
Raimo Kaasila · Sonja Lutovac ·
Pietro Di Martino · Francesca Morselli ·
James A. Middleton · Marilena Pantziara ·
Qiaoping Zhang

Attitudes, Beliefs, Motivation and Identity in Mathematics Education

An Overview of the Field and Future Directions

Gerald A. Goldin
Graduate School of Education
Rutgers University
New Brunswick, NJ
USA

Markku S. Hannula
Department of Teacher Education
University of Helsinki
Helsinki
Finland

Einat Heyd-Metzuyanim
Faculty of Education in Science and Technology
Technion—Israel Institute of Technology
Haifa
Israel

Amanda Jansen
School of Education
University of Delaware
Newark, DE
USA

Raimo Kaasila
Faculty of Education
University of Oulu
Oulu
Finland

Sonja Lutovac
Faculty of Education
University of Oulu
Oulu
Finland

Pietro Di Martino
Dipartimento di Matematica
University of Pisa
Pisa
Italy

Francesca Morselli
Dipartimento Di Matematica
University of Genova
Genoa
Italy

James A. Middleton
School for Engineering of Matter,
   Transport and Energy
Arizona State University
Tempe, AZ
USA

Marilena Pantziara
Cyprus Pedagogical Institute
Nicosia
Cyprus

Qiaoping Zhang
Department of Curriculum and Instruction
The Chinese University of Hong Kong
Hong Kong SAR
China

ISSN 2366-5947   ISSN 2366-5955   (electronic)
ICME-13 Topical Surveys
ISBN 978-3-319-32810-2   ISBN 978-3-319-32811-9   (eBook)
DOI 10.1007/978-3-319-32811-9

Library of Congress Control Number: 2016936641

Printed on acid-free paper

This Springer imprint is published by Springer Nature
The registered company is Springer International Publishing AG Switzerland

Gerald A. Goldin · Markku S. Hannula ·
Einat Heyd-Metzuyanim · Amanda Jansen ·
Raimo Kaasila · Sonja Lutovac ·
Pietro Di Martino · Francesca Morselli ·
James A. Middleton · Marilena Pantziara ·
Qiaoping Zhang

Attitudes, Beliefs, Motivation and Identity in Mathematics Education

An Overview of the Field and Future Directions

 Springer Open

Gerald A. Goldin
Graduate School of Education
Rutgers University
New Brunswick, NJ
USA

Markku S. Hannula
Department of Teacher Education
University of Helsinki
Helsinki
Finland

Einat Heyd-Metzuyanim
Faculty of Education in Science and Technology
Technion—Israel Institute of Technology
Haifa
Israel

Amanda Jansen
School of Education
University of Delaware
Newark, DE
USA

Raimo Kaasila
Faculty of Education
University of Oulu
Oulu
Finland

Sonja Lutovac
Faculty of Education
University of Oulu
Oulu
Finland

Pietro Di Martino
Dipartimento di Matematica
University of Pisa
Pisa
Italy

Francesca Morselli
Dipartimento Di Matematica
University of Genova
Genoa
Italy

James A. Middleton
School for Engineering of Matter,
    Transport and Energy
Arizona State University
Tempe, AZ
USA

Marilena Pantziara
Cyprus Pedagogical Institute
Nicosia
Cyprus

Qiaoping Zhang
Department of Curriculum and Instruction
The Chinese University of Hong Kong
Hong Kong SAR
China

ISSN 2366-5947   ISSN 2366-5955   (electronic)
ICME-13 Topical Surveys
ISBN 978-3-319-32810-2   ISBN 978-3-319-32811-9   (eBook)
DOI 10.1007/978-3-319-32811-9

Library of Congress Control Number: 2016936641

Printed on acid-free paper

This Springer imprint is published by Springer Nature
The registered company is Springer International Publishing AG Switzerland

Contents

Abstract

Research on mathematics-related affect is varied in theories and concepts. In this survey we record the state of the art in this research through short sections from leading experts in different areas. We describe the historical development of the concept of attitude and different ways it is defined. Research on student self-efficacy beliefs in mathematics is summarized. There is reflection on the dialectic relationship between teacher beliefs and practice as well as on how their beliefs change. One section records the emerging research on student and teacher mathematical identities over the last two decades. Finally, mathematical motivation is explored from the perspectives of engagement structures, social behaviors, and the relationship between individual factors and social norms.

Keywords Attitude · Self-efficacy · Beliefs · Identity · Motivation

Attitudes, Beliefs, Motivation, and Identity in Mathematics Education

An Overview of the Field and Future Directions

1 Introduction

Markku S. **Hannula**

The purpose of this publication is to record the current state of the art in research on mathematics-related affect. Research on mathematics-related affect is varied in theories and concepts. Rather than trying to address all perspectives in one chapter, we have identified significant strands of research and invited colleagues from these strands to each write a short section summarizing the state of the art in that strand.

The concepts and theories pertaining to the affective domain can be mapped along three dimensions (Hannula 2012). The first dimension identifies three broad categories of affect: motivation, emotions, and beliefs. In this Topical Survey, motivation is covered in Sect. 2.5 (Middleton, Jansen, and Goldin), which also discusses how emotions and beliefs relate to motivation; Sects. 2.2 (Pantziara) and 1.2.3 (Zhang and Morselli) are on beliefs; and Sect. 2.1 (Di Martino) on attitude more or less cross-cuts through all these categories. The second dimension is movement from rapidly fluctuating state to more stable trait. All of the sections in this chapter focus on trait-type affect while only Sect. 2.5 (by Middleton, Jansen, and Goldin) discusses both of these dimensions (referred to as "in the moment" and "long term"). The last dimension covers the theorizing level, which has three main levels in mathematics-related affect: physiological (embodied), psychological (individual), and social. Mathematics-related affect has mainly been studied using psychological theories and consequently most sections discuss only such research. The so-called social turn (Lerman 2000) in mathematics education is in this Topical Survey mainly reflected in Sect. 2.4 (Heyd-Metzuyanim, Lutovac, and Kaasila) on identity, but Sect. 2.5 (Middleton, Jansen, and Goldin) on motivation also has both a section which discusses the social level and how it interplays with the individual level and a section on self-efficacy which highlights the emerging research on the collective efficacy of collaborative groups. The physiological level

© The Author(s) 2016
G.A. Goldin et al., *Attitudes, Beliefs, Motivation and Identity in Mathematics Education*, ICME-13 Topical Surveys,
DOI 10.1007/978-3-319-32811-9_1

of theorizing is not very popular among mathematics educators, and it is not discussed in any of the sections here. However, it is worth noting the quite extensive neuropsychological research on mathematics anxiety (e.g., Moser et al. 2013; Young et al. 2012).

It is inevitable that such a short publication is not complete. Here I will briefly mention two areas of research that have somehow fallen between the sections and deserve greater attention. One such important area of research on mathematics-related affect is the role of emotions and beliefs in problem solving (Goldin 2000; Hannula 2015). First, emotions such as curiosity, frustration, anxiety, surprise, and elation are an important part of the process of attempting to solve a non-routine problem. Such emotions focus attention and bias cognitive processes. Second, general disposition (e.g., confidence) toward mathematics is known to influence the likelihood of succeeding in any given task.

Gender is another area that deserves a few words. Unlike other research on affect, research on affect and gender "has had a recognized and discernible impact on the development and delivery of mathematics instruction" (Leder and Forgasz 2006, p. 412). Perhaps the most robust research finding in mathematics-related affect is that female students have on average lower self-efficacy in mathematics than male students and similar gender differences tend to also be found in other affective variables (Else-Quest et al. 2010).

Taken together, this summary of research shows the richness of research in this area. There are solid findings that allow the building of theoretical foundations about mathematical affect. At the same time, there are open questions and insufficiently explored venues that call for additional research.

2 Surveys of the State of the Art

2.1 Attitude

Pietro Di Martino

2.1.1 The Pioneering Studies About Attitude: The Measurement Era

In mathematics education, early studies about attitude—a construct developed in the context of social psychology—began to appear in the middle of the 20th century (Dutton 1951). The assumption was that not purely cognitive factors play a role in the learning of mathematics.

In these pioneering studies, the definition of attitude is rarely made explicit, and the main goal is to prove causal correlations between attitude and other significant factors (for example, mathematical achievement). Describing the state of the art, Aiken (1970, p. 592) states, "The major topics covered were: methods of measuring attitudes towards arithmetic and mathematics, the distribution and stability of mathematics attitudes, the effects of attitudes on achievement in mathematics, [and] the relationship of mathematics attitudes to ability and personal factors."

Within this context, researchers follow a quantitative and statistical approach that was considered in that period to be a sort of warranty of the scientific nature of mathematics education. Consequently the focus of the research was mainly the development of new scaling methods (such as Thurstone or Likert Scales or questionnaires) to measure attitude and little attention was paid to theoretical aspects, in particular to the definition of attitude, and to the relationship between attitude and other affective constructs. As Leder (1985, p. 21) underlines: "in many cases, authors either implicitly or explicitly define attitude to mathematics in terms of the instrument(s) used in their research."

At the end of the 1980s, the measurement approach begins to be challenged: several studies show that the correlation between attitude and mathematics achievement is far from being clear. Moreover, the gap between development of instruments and theoretical clarification of the construct began to be considered very problematic, and many scholars explicitly criticized the state, the results, and the *trend* of the research on attitude:

> First, the construct of attitude has been vague, inconsistent, and ambiguous. Second, research has often been conducted without a theoretical model of the relationship of attitude with other variables. Third, the attitude instruments themselves are judged to be immature and inadequate. (Germann 1988, p. 689)

More generally, there has been a gradual affirmation of the interpretive paradigm in mathematics education that has led researchers to try to understand phenomena ("making sense of the world"), abandoning the attempt to explain behavior through measurements or general rules based on a cause-effect scheme (Di Martino and Zan 2015).

As a matter of fact, the shift of perspectives in mathematics education—the movement from a causal-relationship paradigm to an interpretative one—has also deeply influenced research on attitude (Zan et al. 2006) and its methods. The inadequacy of the assumption about cause-effect relationship between attitude and behavior has emerged; attitude is now considered to be an interpretive instrument to understand the reasons for intentional actions: intentional actions involve complex relationships between affective and cognitive aspects; therefore, it is crucial to develop methods able to grasp this complexity.

This shift of perspectives gives new strength to research on attitude that was stuck in the causal-relationship paradigm. In particular, attitude gained renewed popularity in the studies aimed at interpreting the failure in problem-solving activities of students who seem to have the required cognitive resources.

2.1.2 The New Era of Research on Affect (and Attitude)

The beginning of the new era of research on affect in mathematics education can probably be traced to the publication of the book *Affect and Mathematical Problem Solving: A New Perspective* (McLeod and Adams 1989).

This book represents a real turning point in the research on attitude and more generally on affect. Starting with the critique of the research developed on affect until that moment ("This view of beliefs, attitudes, and emotions might be called a black-box approach as opposed to a cognitive approach." Hart 1989, p. 43) and with a shared strong initial assumption ("The initial hypothesis of this project was that affect played an important role in problem solving and that researchers who observed carefully would see the evidence of affect in both students and teachers. That hypothesis has been confirmed." McLeod and Adams 1989, p. 251), the scholars involved in the book highlighted the need to develop a systematic and explicit theoretical framework for dealing with affect (useful to interpret the relationship among affective constructs and between them and cognition). In particular, the need to clearly define the constructs and develop coherent methodologies is stressed:

> There was a lack of definition, lack of clarity, and lack of connections to mathematics. It is possible to avoid making the same mistakes again as new ideas and research methodologies are employed. It is hoped that new researchers on affect will be clear about what is being studied, precise in definition, and respectful of what has been learned previously. (Fennema 1989, p. 209)

A few years after the publication of *Affect and Mathematical Problem Solving*, based on the needs stressed in the book, McLeod (1992) proposed a new framework for research on affect in mathematics education. He identifies three main constructs (emotions, beliefs, and attitudes) and characterizes them. But, as Hannula (2011) underlines:

> Probably the most problematic concept in McLeod's framework is attitudes. Within mathematics attitude research, attitudes have typically been defined as consisting of cognitive (beliefs), affective (emotions), and conative (behavior) dimensions. If we try to combine the tripartite framework with McLeod's, we see that attitude is at the same time a parent and a sibling to emotions and beliefs. (p. 38)

As a matter of fact, in those years researchers provided a variety of definitions of the concept of attitude: all of them involve other factors. In particular, two definitions of attitude are particularly recurrent: a *simple* definition that describes attitude in terms of positive or negative feelings associated with math and a *three-dimensional* definition that recognizes three components in attitude (the emotional disposition, the set of beliefs regarding mathematics, and the behavior related to mathematics). Both the two definitions show enormous theoretical limits (Di Martino and Zan 2001).

The debate about the several definitions of attitude led researchers to consider the *suitableness* of the definition rather than its *correctness*: the adequacy of the definition depends on the issues studied. This was fundamentally the idea of Daskalogianni and Simpson (2000): they suggested considering the definition of attitude to be a working definition: a function of the problems that the researchers pose themselves.

This kind of approach characterizes the new trend of research on attitude as *problem-led*. This view is in line with the very interesting position of Ruffel et al. (1998): "we conjecture that perhaps it [attitude] is not a quality of an individual but rather a construct of an observer's desire to formulate a story to account for observations" (p. 1).

In relation to the discussion about definition, scholars have debated about the adequacy of methods in research about attitude and their coherence with the definition used.

Within the new interpretative paradigm, the development and use of qualitative methods for research on attitude emerges. In particular, much of the research about attitude has been developed through narratives such as essays, diaries, and interviews (Karsenty and Vinner 2000; Hannula 2002; Kaasila 2007; Di Martino and Zan 2011).

The main strength of this narrative approach that has clearly emerged is the possibility of collecting the aspects and details that respondents consider relevant in the development of their relationship with mathematics. The narrative approach differs from the use of traditional attitude scales—where respondents are requested to express agreement/disagreement on items chosen by others that are sometimes irrelevant for them—in that respondents can specify what they consider crucial and skip what they consider irrelevant. That is, the narrative approach brings out what is central for the respondents.

2.1.3 The TMA Model for Attitude: A Characterization of Attitude Grounded in School Practice

The attitude construct has been widely used by mathematics teachers: often teachers' diagnosis of "negative attitude" is a causal attribution to students' failure and perceived as global and uncontrollable rather than an accurate interpretation of students' behavior that is capable of steering future action. To make this diagnosis useful for dealing with students' difficulties in mathematics, we conducted a long study based on the collection and analysis of students' autobiographical narratives (Di Martino and Zan 2011) in order to construct a characterization of attitude strictly linked to students' experience with mathematics.

An analysis of 1662 anonymous essays entitled "Maths and me: my relationship with maths up to now" written by students of all school levels was conducted. According to a grounded-theory approach to the data (Glaser and Strauss 1967), we used the collected data to discover a set of categories aimed at understanding how students describe their own relationship to mathematics.

At the end of our analysis, we identified three main dimensions in students' narratives: emotional dispositions towards mathematics, view of mathematics, and perceived competence in mathematics (only 32 essays, 2.1 % of the entire sample, did not refer to at least one of these three dimensions).

Fig. 1 The TMA model

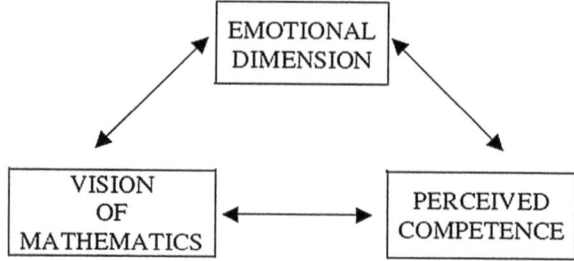

Therefore we propose a three-dimensional model for attitude (TMA) characterized by the three dimensions that students recognize as crucial in the development of their relationship with mathematics and by their mutual relationships (see Fig. 1).

The arrows in the schema have a crucial role: TMA takes into account the relationship among the three dimensions. These relationships appeared clearly in the students' narratives.

The subjectivity of these relationships among the three dimensions that emerged as one of the results of our research confirms the complexity of the construct:

> The proposed model of attitude acts as a *bridge* [italics in the original] between beliefs and emotions, in that it explicitly takes into account beliefs (about self and mathematics) and emotions, and also the interplay between them. However, in order for it to become an effective theoretical and didactical instrument, the construction and use of consistent instruments for observation, capable of taking into account its complexity, are needed. (Di Martino and Zan 2011, p. 479)

Through the TMA model we have interpreted some recurrent phenomena in the development of attitudes towards mathematics and above all we have given a more sophisticated definition of negative attitude (Di Martino and Zan 2010).

In particular, we have identified different profiles of negative attitude, suggesting implications for teacher practice and for teacher education in order to overcome what we have called the *black box approach*: "that student has a negative attitude toward mathematics" is often the teacher's claim of surrender rather than a precise diagnosis to activate a didactical intervention.

We have briefly described the narrative of the research on attitude towards math (for a more complete report see Di Martino and Zan 2015). This narration must surely be continued—the debate about some critical issues still continues and new issues and new goals have emerged (see Looking Ahead section)—but it is a fact that in the last 25 years the research on attitude in mathematics education has moved ahead in important ways: overcoming a naive approach to the construct, discussing methods, and producing some solid findings (Zan 2013). These solid findings are the significant heritage for those who follow.

2.2 Student Self-efficacy Beliefs

Marilena Pantziara

2.2.1 Introduction

Bandura (1997) defined perceived self-efficacy as "beliefs in one's capabilities to organize and execute the courses of action required to produce given attainments" (p. 3). Self-efficacy tends to be conceptualized as a context-specific evaluation of one's competence to perform a specific task and it differs conceptually and psychometrically from other related beliefs such as outcome expectations, self-concept, and perceived control (Pajares 1996; Williams and Williams 2010). Since Albert Bandura's publication of the theory of self-efficacy in 1977, researchers in every domain of social sciences have explored the influence of self-efficacy in people's behavior, with the mathematics domain being one of the most investigated (Pajares 1996).

Self-efficacy indices focus on cognitive beliefs which are created and altered through the interpretation one makes during four types of experiences. Mastery experiences are the most influential sources since they are predicated on the outcomes of personal experiences. In educational settings previous success develops students' self-efficacy while failure undermines it. Vicarious experiences depend on an observer's appraisals of capabilities in relation to others and outcomes attained by a model. Verbal persuasion has a more limited impact on students' self-efficacy since outcomes are described and not directly experienced. Last, students' psychological reactions such as stress, tiredness, and other emotions are often interpreted as indicators of physical incapability (Bandura 1986; Zimmerman 2000).

The information from these four sources is only helpful if people cognitively interpret them. While the evaluation of the sources of self-efficacy is not well understood, it is believed that the process involves interpretation in which individuals choose and weigh selected factors. As Klassen and Usher (2010) state, there is no one-size-fits-all approach where self-efficacy can be developed by providing prescribed sources of self-efficacy, since each individual interprets the information in a unique way. In this context, *reciprocal determinism* is the term used by Bandura to describe the continuous reciprocal interaction between behavioral, cognitive, and environmental influences (Williams and Williams 2010). In this process, individuals are agents, able to serve both as results and as makers of their own environment.

In mathematics education, students working in small groups is a mandate in every reformed curriculum. Students working in groups develop, apart from their own beliefs about their capabilities, shared beliefs about their capabilities as a group. Collective efficacy is defined by Bandura (as cited in Klassen and Krawchuk 2009) as "a group's shared belief in its conjoint capabilities to organize and execute the courses of action required to produce given levels of attainment" (p. 102). While self-efficacy and collective efficacy beliefs are closely related, collective efficacy is more than the sum of the combined group self-efficacy since it is developed when a group works together.

2.2.2 Student Mathematics Self-efficacy Beliefs
and Unresolved Issues

Research stretching back to the mid-1980s, mostly in the realm of educational psychology, has highlighted the ways in which self-efficacy beliefs influence students, teachers, and group attainments in many domains, mostly in mathematics. Generally, studies have shown that self-efficacy operates directly and indirectly on behavior. Specifically, studies related to students and self-efficacy have focused on two major areas. One area explored the link between self-efficacy, course selection, and career choices. Studies in the second area have investigated the relationship between self-efficacy, different psychological constructs, academic motivation, and achievement. Specifically, students' self-efficacy has been found to be a significant predictor of students' course selections, academic continuance and achievement, and college performance and achievement. Academic self-efficacy has been found to influence students' persistence, self-regulated strategies, and effort. Students' self-efficacy has been found to influence them emotionally by decreasing their stress, anxiety, and depression (Klassen and Usher 2010; Pajares 1996; Zimmerman 2000).

In their meta-analysis on self-efficacy beliefs, Multon et al. (1991) revealed that self-efficacy was positively correlated to academic outcomes and persistence outcomes across a wide variety of subjects, experimental designs, and assessment methods. An important finding was that a factor moderating this relationship involved subject age. Specifically, among students in the normal achievement range, high school and college students evidenced stronger effect size than did elementary school students. The researchers argued that older students that possess greater school experience and more well-defined perceptions about their strengths and weaknesses can make more accurate self-efficacy judgements.

In all of these studies, while there is a clear relation between self-efficacy and the different constructs, effect sizes and relationships varied depending on the way in which self-efficacy was measured and assessed. Reviewers on self-efficacy research note that there are persistent difficulties in self-efficacy research (Klassen and Usher 2010). Particularly, Pajares (1996) refers to the mismeasurement of self-efficacy, noting that because judgements of self-efficacy are task and domain specific, global or inappropriately defined self-efficacy evaluation weakens effects. He warned that in order to avoid an atheoretical measurement of broad attitudes about general capabilities with a passive resemblance to self-efficacy, researchers should follow theoretical guidelines regarding specificity of self-efficacy assessment and correspondence with criterial tasks.

As far as it concerns self-efficacy and mathematics achievement, although these studies show that self-efficacy and achievement are clearly related, very few provide causal ordering between the two constructs (Skaalvik and Skaalvik 2011). Moreover, while reciprocal determinism is well endorsed in the literature, there is a lack of empirical support, specifically referring to the reciprocal influence of self-efficacy and mathematics performance. Williams and Williams (2010) note that this is due to the relative availability of data believed to be necessary for this examination, specifically data with repeated measures separated by time, allowing for the estimation of cross-lagged effects.

Empirical support is sparse also in longitudinal studies. In particular, most longitudinal studies to date have included limited ranges (e.g., middle childhood or adolescence). In addition, little research has examined the specific relation between self-beliefs and their associated behaviors across time (Davis-Kean et al. 2008). Last, comparative studies focusing on the relationship between self-efficacy and cognitive variables in mathematics education are also rare.

To date, most research studies on the sources of self-efficacy have been in the field of mathematics education. In their review of the literature, Usher and Pajares (2008) refer to quantitative and qualitative ways of measuring sources of self-efficacy. The adapted version of the Sources of Mathematics Self-efficacy Scaled developed by Lent (1991, as cited in Usher and Pajares 2008) and his colleagues has been used in many studies. The findings of various studies show that mastery experience consistently emerges as the most powerful source of self-efficacy, while findings for the other three sources have been less consistent. Usher and Pajares (2009) note that these inconsistent results may be due to methodological problems such as poor reliability, aggregated scores that mask information from any one source, or multicollinearity between the sources. Usher and Pajares (2008) state that the sources of self-efficacy function best at appropriate levels of specificity and when they correspond with the self-efficacy outcome they are developed to predict.

The research related to collective efficacy has been found to be significantly correlated with group performance and other collective motivation constructs such as group cohesion (Klassen and Krawchuk 2009). Despite the recognized importance of collective efficacy, almost all research is focused on teachers' collective efficacy beliefs and very little attention has been paid to the collective efficacy of students in mathematics.

2.2.3 Recent Developments in Student Mathematics Self-efficacy Beliefs

The unresolved issue of the measurement diversity of self-efficacy beliefs as it has been described by Pajares (1996) and many years later by Klassen and Usher (2010) seems to occur also in recent studies related to self-efficacy. In addition, there have recently been several pleas to further develop the available educational and psychological measuring instruments by taking into account subject-specific aspects and to study student affect in a domain-specific, subject-oriented way, particularly by using tasks (Schukajlow et al. 2012).

Aligned with these appeals, a recent instrument was developed by Pampaka et al. (2011) that was designed to measure mathematics self-efficacy as a learning outcome of students following post-compulsory mathematics programs. The instrument was applied to 1779 students and the results showed one possible misfit to the model and led to the hypothesis that there may a need for two sub-dimensions in the construct of mathematics self-efficacy beliefs: pure and applied.

Recent development in the study of student self-efficacy and mathematics shows the focus to be on (a) longitudinal studies, (b) comparative studies, (c) experimental designs, (d) studies related to sources of self-efficacy beliefs, and (e) collective self-efficacy. Some of these recent studies are presented below.

Hannula et al. (2014), in a longitudinal study with 3502 Finnish students from the beginning of Grade 3 to the end of Grade 9, investigated the direction of causality between mathematics-related affect and achievement. An important finding was that students' mathematical achievement, emotion, and self-efficacy were significantly stable over time. The results indicated that mathematics achievement and self-efficacy have a reciprocal relation where the dominant effect of this relation is from achievement to self-efficacy.

Skaalvik and Skaalvik (2011) in two longitudinal studies examined whether mathematics self-perception (self-concept and self-efficacy) predicted achievement over and above the prediction that could be made by prior achievement. The participants were 246 middle school students and 484 high school students. The analysis showed that students' self-efficacy strongly predicted achievement over and above the prediction that could be made from prior achievement and that self-concept and self-efficacy were important mediators of academic achievement.

Davis-Kean et al. (2008) examined the relation between self-efficacy beliefs and behaviors across time to determine the stability of this relation. The relation between beliefs about self and behavior was examined in two independent data sets with two different constructs: aggression and achievement. The sample consisted of students aged 6–18 years old (Grades 1–12). The results revealed that self-beliefs become more strongly related to behavior as students grow older. These findings were replicated in four independent samples. The results showed that this relation did not differ by gender. The authors argued that the early development of students' beliefs is a precursor to later knowledge about the self.

Self-efficacy beliefs are considered to be shaped by the sociocultural context in which they develop. Much of the investigation of self-efficacy beliefs and performance has developed in Western societies. The question of whether the self-efficacy construct has the same meaning in different contextual settings is important to consider.

The comparative study by Williams and Williams (2010) investigated the reciprocal determinism between self-efficacy and performance in 33 nations on the basis of the PISA 2003 data. The reciprocal determinism of mathematics self-efficacy and achievement was supported in 24 of the 33 nations.

Despite the ongoing debate about causal relation between self-efficacy and mathematics achievement, there is a general agreement that self-efficacy beliefs are formed through experience in the environment (Skaalvik and Skaalvik 2011). In a recent study by Schukajlow et al. (2012), 224 ninth grade German students were asked about their enjoyment, interest, value, and self-efficacy expectations concerning three types of mathematical problems: intra-mathematical problems, word problems, and modelling problems. Half of the students received student-centered teaching while the other half teacher-centered teaching. The findings indicated that there were no differences in students' enjoyment, interest, value, and self-efficacy between the three types of problems. A comparison of post- and pre-test means showed that the specific teaching unit of modelling problems had a positive effect on students' affect for the three types of problems addressed. Students' enjoyment, interest, and self-efficacy beliefs increased significantly, with the student-centered teaching method producing the most beneficial effects.

Studies suggest several instructional practices that could enhance student self-efficacy. Özdemir and Pape (2013), focusing on two sources of efficacy, mastery experiences and social persuasion, examined how these sources were structured for three students with different levels of mathematics achievement and self-efficacy within a sixth grade mathematics classroom. The results for each case showed that each student experienced success and received social persuasion differently.

Last, one of the very rare studies on collective efficacy of students conducted by Klassen and Krawchuk (2009) examined the collective efficacy in 125 randomly assigned groups of older (mean age 13.45 years) and younger (mean age 11.41 years) students. The students completed three cooperative small-group tasks involving puzzles and mathematics operations along with individually completed measures of self-efficacy, collective efficacy, and group cohesion. For the older students, groups with high collective efficacy and group cohesion scored higher on performance tasks than groups with low collective efficacy and group cohesion. An important finding was that collective motivational beliefs of the older students were more closely related to performance as the groups worked together over time on the tasks.

Having presented some of the trends that have emerged the last decade concerning student self-efficacy in mathematics, we may conclude that there is a move towards multi-method approaches to understand the concept and to identify relations among this area of research. Recent studies in the field move beyond survey research to longitudinal and experimental approaches. Self-efficacy in mathematics is gaining increased interest internationally, but there is still much to be clarified and revealed on the basis of high quality measurement.

2.3   Teacher Beliefs

Qiaoping Zhang and Francesca Morselli

2.3.1   Introduction

More than 30 years ago, Fenstermacher (1979) predicted that the study of beliefs would become the focus for teacher effectiveness research. Since the 1990s, research has begun to focus on the affective factors behind teacher teaching behavior, particularly on teacher beliefs. In the section below we will give a holistic review of empirical and theoretical studies on teacher beliefs performed in more recent years.

As many researchers have pointed out, there is no internationally accepted definition about beliefs. Thompson (1992) remarked that "for the most part, researchers have assumed that readers know what beliefs are" (p. 129). Furinghetti and Pehkonen (2002) emphasized the subjective and hidden characteristics of beliefs. Philipp (2007) proposed that "beliefs might be thought of as lenses that affect one's view of some aspect of the world or as dispositions toward action" (p. 259). Though there are many categorizations in literature, in general teacher beliefs about mathematics ranged from viewing mathematics as a static, proce-dure-driven body of facts and formulas to a dynamic domain of knowledge based on sense making and pattern seeking (Ernest 1989; Cross 2009).

Teacher beliefs seem to originate from personal experience, experience with schooling and instruction, and experience with formal knowledge (Richardson 1996). Moreover, beliefs are not to be regarded in isolation, rather they are in clus-ters and constitute a belief system (Rokeach 1968; Green 1971).

2.3.2 What Has Been Done in Recent Years

A main trend of research addresses the *dialectic relationship* between beliefs and practice (or, similarly, between professed beliefs and beliefs inferred by the observed practice). However, this relationship cannot simply be viewed as linear. Both consistencies and inconsistencies have been found between teacher beliefs and their practices (Cross 2009; Wilkins 2008). From a methodological point of view, Speer (2005) argued that the divide between teacher professed beliefs and attributed beliefs (inferred from observation of practice) is only an apparent dichotomy: indeed, the divide may be interpreted in terms of a mismatch between teachers and researchers of the meaning of the terms used to describe beliefs and practices. Schoenfeld (2011) also pointed out that the literature on beliefs, spe-cifically in teaching, has been largely descriptive, often focusing on teachers' pro-fessed beliefs. What matters in teaching is not so much what people say but what they do. Factors such as knowledge, experience, goals, and context, which can shape teacher beliefs, need to be investigated further.

Concerning the supposed *inconsistency* between attributed beliefs and practice, Leatham (2006) reminded researchers that they must look deeper "for we must have either misunderstood the implications of that belief, or some other belief took precedence in that particular situation" (p. 95). We should regard beliefs as sen-sible systems. Two works may be inserted in this stream of research. Furinghetti and Morselli (2011), in their study of teachers' practice concerning proof, focused on the detection of the reasons behind teachers' decisions. They proposed the construct of leading beliefs, i.e., "beliefs (whose nature may vary from teacher to teacher) that seem to drive the way the teacher treats proof" (p. 590). In the same vein, Cross (2015) focused on supposed inconsistencies between beliefs and prac-tice "to better understand the broader set of beliefs that could be influential in the teacher decision making and behavior" (p. 191). She pointed out the crucial role of other general beliefs and contextual factors on teacher classroom behavior.

As Fang (1996) suggested, the reasons for inconsistencies might be the com-plexities of classroom life. Teacher beliefs are situational and are manifested in instructional practices only in relation to the complexities of the classroom context (Wong et al. 2016). The focus on *context* may be traced back to the influential work by Skott (2009), who advocated a "social turn" in the research on teacher beliefs. This means a shift from a study on beliefs seen as "explanatory principles" for classroom practice to a locally social approach, where context plays a very crucial role. In other terms, classroom practice is not determined by beliefs, but emerges in and though interaction with context. As a result, a supposed conflict between espoused and enacted beliefs should rather be read as a conflict between espoused beliefs and the ways of acting within the context. This means that research should focus on teachers' contexts and the actual and virtual communities of practice teachers live in, not only on beliefs. Beswick and her colleague's work (2012, 2014) on different persons' beliefs provided another perspective on context.

Another trend of research addresses the issue of teacher belief *change*. Many studies have declared stability to be an important feature of beliefs and have dealt with belief change without any initial measure of beliefs. Liljedahl et al. (2012) argued that stability should not be seen as a defining quality of beliefs, rather the result of some (but not all) empirical research on beliefs. There are several examples of research studies aimed at promoting belief change in the context of teacher education programs. For example, Swan (2007) discussed the impact of tasks on teachers' beliefs and practice. He argued that the effect occurred in two ways: beliefs affected the task implementation and, conversely, the task implementation might affect beliefs. Charalambous et al. (2009) studied the use of history of mathematics as a means to change pre-service primary teachers' beliefs about mathematics. Grootenboer (2008) pointed out the crucial role of teacher education programs in helping teachers to reflect on their existing beliefs and pointed at some ethical issues concerning the role of the teacher educator in situations where there was a shift from encouraging prospective teachers to reflect on existing beliefs to promoting a change in beliefs.

Beliefs may change in a rapid and profound way. Liljedahl (2010) listed a series of "types" of belief change, each underlying a different transformation: (1) conceptual change, (2) accommodating outliers, (3) reification, (4) leading belief change, and (5) push-pull rhythm of change. Among them, we refer to leading belief change (which may be connected to the aforementioned studies on leading beliefs affecting practice) and conceptual change. Conceptual change happens when an existing belief starts to be questioned or even rejected by a teacher; such a change is profound when the teacher finds a new belief to replace the former one. In a subsequent work, Liljedahl (2011) deepened the issue of teacher change as conceptual change, and argued that the theory of conceptual change may act as a theory *for* changing beliefs, that is to say for planning teacher development interventions.

Most studies address beliefs about mathematics teaching and learning without reference to any specific mathematical domain, process, or topic. Some exceptions concern the teaching and learning of proof (Furinghetti and Morselli 2009), the use of multiple solution tasks (Guberman and Leikin 2013), the teaching of calculus (Erens and Eichler 2014), and problem solving (Andrews and Xenofontos 2015).

2.4 Identity

Einat Heyd-Metzuyanim, Sonja Lutovac and Raimo Kaasila

2.4.1 Introduction

The last two decades have seen a significant increase in studies that focus on identity in mathematics education. These studies have largely been a part of the "social turn" (Lerman 2000) in the field, where increasing attention has been given to the social, cultural, and political aspects of mathematics teaching and learning. Broadly speaking, the studies in this domain can be divided into two major categories: student identity and teacher identity. These two bodies of literature have developed independently, though they often share similar theoretical backgrounds. Therefore, we shall give an overview of them in two separate sections, while pointing to their similarities and differences in the summary part.

The present chapter of this Topical Survey aims to give a brief general overview. In preparation for this overview, we collected and reviewed all the peer-reviewed journal papers and books that we could find that included "identity," "identities," and "mathematics" either in their title, abstract, or keywords. The reference list can be obtained upon request.

In our overview, we intended to answer the following questions:

1. What have been the theoretical frameworks used for examining student and teacher identities, how is identity defined, and what are the methods used for its study?
2. What are the major findings in this literature?
3. What seems to be missing? What are further avenues for research that seem to be important or productive?

2.4.2 Student Identity

Almost all the studies on identity have stemmed from socio-cultural theories of learning. Prominent in these frameworks are socio-linguistic theories (Gee 2001), Wenger's learning-in-participation theory (1998), Holland's framework of "figured worlds" (Holland et al. 1998), positioning theory (Harré and van Langenhove 1999), socio-political theories, cultural historical activity theory (CHAT), critical race theory, and Sfard's (2008) "commognitive" framework. Many of the writings make use of some combination of the above theories. Psychological theories of identity development have been less important, though Erikson's influential theory is mentioned at times. The main theoretical link that has been made between identity and learning is through the concept of participation. Students have been theorized to not only acquire knowledge but also become a certain person through learning mathematics. Thus developing and identity of inclusion (or exclusion) in the community of mathematical learners has been the focus of many of these studies (e.g., Solomon 2009).

One of the early pitfalls of the study of mathematical identity was lack of clarity in definitions of this term. An important step forward was made by Sfard and Prusak (2005), who proposed to define identity as "collections of stories about persons … that are reifying, endorsable, and significant" (p. 16). In later publications, many writers defined identity more explicitly, though often authors resisted constraining their definition to narrative and included in it "beliefs," "perceptions of self," "perceptions of mathematics," and "ways of being" (e.g., Bishop 2012). In particular, authors have stressed the importance of performances or enacted identities (Varelas et al. 2013). In addition, many authors have refined the concepts related to identity, including "current and designated" identities (Sfard and Prusak 2005), "normative" and "personal" identities (Cobb et al. 2009), "leading identities" (Black et al. 2009b), and "identifying" as a process of identity construction (Heyd-Metzuyanim and Sfard 2012).

Methodologically, the vast majority of studies on student identity rely on qualitative tools. Very few (such as Bishop 2012) have made use of quantitative measures, mainly via coding and counting of utterances. Though most of the studies rely on interviews, some have shifted attention to the ways in which student identities are enacted and co-constructed in the activity of learning. Bishop (2012), Wood and Kalinec (2012) and Heyd-Metzuyanim's (Heyd-Metzuyanim and Sfard 2012; Heyd-Metzuyanim 2015) studies have developed methodologies for examining the moment-to-moment interactions by which students identify themselves and others as competent or incompetent in mathematics. They showed that these identification processes had a significant impact on the process of learning and on opportunities that students were given or took up during learning activities.

Regarding findings, these are often unique to the specific situation under examination. Still, some major influential findings can be cited. One of them is Boaler's (e.g., Boaler and Greeno 2000), who showed that students studying in different learning environments, namely lecture-based versus problem-solving and group-work based learning, developed different mathematical identities. Students in problem-solving settings developed identities of competence that included mathematics as part of their envisioned future, while students in lecture-based settings did not. Cobb and his colleagues (e.g., Cobb et al. 2009) corroborated these findings by showing how students in a problem-solving setting aligned themselves to the new norms of participation in ways that could positively influence their mathematical identity. Linking the issue of traditional learning settings to issues of race and gender, Lim (2008) showed that self-identity narratives of three sixth grade girls were tightly linked to their narratives about race, gender, and social class and that this interaction was to the disadvantage of the African-American girl in the study.

In general, the issue of race and gender has figured prominently in this literature. Martin (2007) showed the conflicts experienced by black African-American boys between their race and culture and the narratives of being a good mathematics student. Nasir (2002) reported how out-of-school activities that are popular within African-American youth (e.g., dominoes and basketball) are experienced as disconnected from school mathematics although they include significant mathematics. Oppland-Cordell and Martin (2014) extended this line of research to Latin@s' mathematical identity. Research intersecting identity and race has been carried on mainly in U.S. settings (though some work in PME conference proceedings, not covered in this review, has also been relating to racial issues outside the United States). Research in Europe, on the other hand, has tended to relate to identity in more general terms, as located in socio-cultural and socio-political spaces but not necessarily connected to a specific racial or ethnic group (e.g., Andersson et al. 2015; Black et al. 2009a).

2.4.3 Teacher Identity

After the 2000's, several lines of research on mathematics-related teacher identity emerged, including the samples of: (a) pre-service elementary teachers (Hodgen and Askew 2007; Jones et al. 2000; Lutovac and Kaasila 2011; Ma and Singer-Gabella 2011; Walshaw 2004), (b) pre-service mathematics teachers (de Freitas 2008; Goos 2005; Goos and Bennison 2008), (c) in-service elementary teachers (Spillane 2000; Drake et al. 2001), (d) in-service mathematics teachers (Graven 2005; Hodges and Cady 2012; van Zoest and Bohl 2005), and (e) teacher educators (e.g., Grootenboer 2013). There is also a relatively new line of research addressing the identities of mathematics coaches (Chval et al. 2010).

The definitions of teacher identity in these studies usually go hand in hand with the theoretical approaches of the concept. One of the most widely used is socio-cultural, building on Lave and Wenger's (1991) and Wenger's (1998) work, seeing identity as a way of belonging to different communities of practice and an activity of participating in them (e.g., Goos 2005). Other perspectives include post-structural (Brown and McNamara 2011; de Freitas 2008; Walshaw 2004) and psychoanalytic (e.g., Black et al. 2009a), as well as perspectives deriving from multiple areas of research. This situation has resulted in a variety of definitions and often the absence of them altogether. In terms of methodology, most studies are conducted as small-scale qualitative studies, and there seems to be an expansion in the use of narrative and discursive methods.

Generally, teacher identity is seen as a dynamic construct, i.e., changing over time and with a general consensus on its contextuality. Thus, studies have tended towards subject-matter identities such as that of a mathematics teacher, but also towards multiple identities bound to diverse teacher roles and the multiple communities they participate in. Studies have also pointed to the fact that the nature of teacher identity may be different in different countries (Leung 2001; Lutovac and Kaasila 2014). Most studies in one way or another demonstrate the interaction between teacher identities and teachers' practices, making it apparent that changing one will affect the other. Moreover, the external demands posed on teachers (e.g., school reforms) inevitably affect teachers' identities: teachers often see changes as threatening to their identity, thus their identity becomes an obstacle for change (see also Gellert et al. 2013).

The research has highlighted the process of becoming, often situated in teachers' stories or narratives and associated with teachers' prior experiences (e.g., Hossain et al. 2013; Lutovac and Kaasila 2011, 2014; Neumayer-Depiper 2013). These findings suggest that teachers' personal histories, such as those of being a learner, undoubtedly shape and become a part of their teacher identities. Many of the studies promoted an awareness of the need to implement teacher identity construction in teacher preparation programs, such as in mathematics education courses, via online communities of practice and ICT, and particularly in teaching practicums (da Ponte et al. 2002; Goos and Bennison 2008; Walshaw 2004). In addition, emotions are highlighted in the process of learning to become a teacher of mathematics (e.g., Hodgen and Askew 2007). The effort to address the so-called affective component of identity construction has been done especially in relation to pre-service elementary teachers, who were shown to experience great difficulties with the subject. There is some evidence that encouraging pre-service teachers to narrate their own or listen to their peers' personal experiences with the subject makes them cope better, which may lead to the development of a more suitable identity for mathematics teaching (Kaasila et al. 2008; Lutovac and Kaasila 2011, 2014).

2.5 Motivation

James A. Middleton, Amanda Jansen and Gerald A. Goldin

2.5.1 Introduction

Mathematics education has been plagued, over the years, by a kind of paradox. We teach mathematics for the public good, with the goal that all citizens be able to reason quantitatively; understand scientific, economic, and social arguments based on data; and use this understanding to make informed decisions about themselves and our collective polity. Yet in many countries, while mathematics is viewed as beneficial societally, it is not seen by a majority of students as beneficial personally. Such beliefs and norms about mathematics have hindered significant progress in democratizing access to quality mathematics teaching and learning, and even those with such access tend to avoid advanced mathematical topics and courses (Simpkins et al. 2006). This paradox is also manifest in public attitudes towards science and other mathematically intensive fields (National Science Board 2014).

In our view, the root of the problem lies not in mathematics content per se. We do not regard mathematics as more difficult, more complex, or more boring in and of itself than other academic content. Rather, the norms, beliefs, and practices that have arisen over the past century and a half related to mathematics teaching, learning, and assessment, have ignored or poorly articulated the role of motivational processes in mathematics learning.

This section considers the broad field of motivation in mathematics education research. We suggest that, until recently, the focus on individual motivational processes to the exclusion of social norms and practices has prevented our field from developing coherent and effective theories of instruction related to student motivation. Correcting this theoretical deficit requires a different approach to the study of motivation. We propose consideration of affective structures, which are highly social in their very nature; a focus on helping behaviors, of which teaching practices are only a small subset; and incorporating explicitly the relationship between individual autonomy and the social norms that bound it.

2.5.2 Engagement

We have suggested (Middleton et al. in press) that the focus of motivation research be shifted from the study of longer-term attitudes and beliefs toward that of in-the-moment engagement. Briefly, mathematics engagement involves the simultaneous recruitment of motivational and affective structures to guide sustained, productive learning behavior. Critical to this perspective is the idea that "productive learning behavior" is a social construct formed from the interaction of learners' personal learning states and mathematical dispositions, their home community, their classroom or learning environment community, and macro-cultural constraints such as curriculum, assessment, and cultural attitudes. Such interaction creates *structures* of engagement that are relatively stable under certain eliciting conditions. Examples include: Get the Job Done, where the primary motivation is social, a deference to the teacher, parent, or another person or persons with whom the student feels allied; I'm Really Into This, where the primary motivation is intrinsic and doing challenging mathematics is experienced as its own reward; or Check This Out, where performing successfully in mathematics is motivated primarily by an extrinsic reward or a perception of utility (Goldin et al. 2011). The point here is that a combination of intrinsic, extrinsic, social, and individual factors are interrelating whenever a student engages in mathematical activity. Paying attention to the interactions among these factors can help us identify structures of engagement and their eliciting conditions, reinforcers, and social constraints and perhaps develop catalytic strategies and tools by which teachers can improve engagement for more students in more challenging mathematics.

2.5.3 Motivation and Self-regulation in Mathematics

Motivation is, put simply, the reason we engage in any pursuit, mathematical or otherwise. Human beings have interests, goals, and preferences, and these structures serve as templates for whether to put forth effort towards mathematical activity and the extent to which efforts are seen as efficacious. Moreover, while engaged, these structures help us monitor and direct our efforts towards resolution of the goals we have set for our engagement, including the recruitment of cognitive and affective resources that improve our chances for success. Because of this, motivation is central to self-regulation (Boekaerts et al. 2005).

Zimmerman (2005) presents self-regulation as an amalgam of cognitive and motivational evidence that shows us that individuals are able to adapt—i.e., regulate—their thoughts, behaviors, and environmental conditions to fulfill personal goals. He and other researchers (Middleton and Toluk 1999) have posited a cyclic process of Anticipation → Engagement → Reflection that enables learners to prepare for engagement, control their behavior while engaged, and look back and assess what went well and what fell short so that they are better prepared for similar tasks in the future (see Fig. 2). In this cycle, when confronted with a mathematical activity, students first do a kind of task analysis of the potential activity to determine the value the task has, to choose whether or not to engage, and to plan their course of action. Second, they choose to engage and recruit strategies and regulate their performance. Third, they evaluate their performance, store memories of successful and unsuccessful strategies for later recall, and assess the value (i.e., interest, reward, etc.) of their involvement (Schunk and Zimmerman 1998).

Understanding motivation as a regulatory process is critical, we think, because there is tension in the research literature between the here and now of task engagement, and the longer-term patterns of engagement we see in mathematics learners. In the here and now, we can see students developing interest, and engaging with gusto, while over the long term, they may try to avoid further mathematics coursework. Or, students may be bored with a particular algebra task, but tend to enjoy and seek out algebraic puzzles in their free time. What we see is the cycle of self-regulation that serves to manage engagement behaviors, but may or may not result in long-term valuing of mathematics, positive affect, and improved mathematics performance. When experiences tend to be consistent and coherent over time with regard to motivational affordances, it becomes more probable that the person will develop a (positive or negative) long-term disposition and identity toward mathematics.

Our affective responses serve both informational and reward functions in this process: positive affect is usually an indicator that the strategy we have chosen has been effective, and it feels good as well! Negative affect also encodes information: e.g., that our efforts have been ineffective, or that the task has little value. In some cases, though, negative affect can be a springboard for renewed effort, when successful outcomes are highly valued by the learner (Frenzel et al. 2007).

Fig. 2 In-the-moment self-regulation cycle. Adapted from Schunk and Zimmerman (1998)

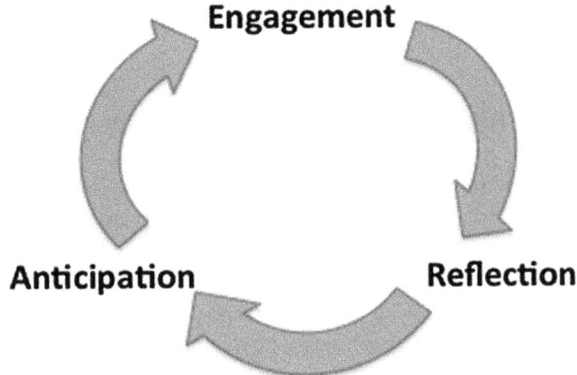

Behavior is primarily regulated by social norms and practices, and constrained by cultural attitudes, available tools, and materials. Self-regulation typically works (productively) within these constraints and affordances. Moreover, the dynamics of actual mathematical tasks, particularly those performed in small cooperative groups, necessitate re-thinking the traditional self-regulation triad.

2.5.4 Self-determination Versus Social Determination

There is a tension in the literature (and in classrooms) by which student engagement is considered either autonomous or social. This is a false dichotomy, of course. People do choose among both intrinsic motivation and the extrinsic motivators to which they ascribe value, and integrate these factors into their own identity (Deci and Ryan 2008). Thus, when students are motivated autonomously, they have an internal register of worthwhile activities, and a repertoire of strategies and behaviors that they can draw upon to be successful. But they also rely on social cues and are constrained by the norms of classroom behavior, including social goals, general class affect, and more specific relationships with other students (Webel 2013). Even the most autonomously directed mathematical behavior is guided by social rules and relationships.

Moreover, when students exhibit controlled motivation, consisting of purely external reinforcement contingencies such as treats and threats or introjected regulation where such contingencies have been partially internalized through association with other more valued outcomes (such as grades indicating approval), they experience social pressure to act and emote in socially appropriate ways (Deci and Ryan 2008). However, they also have some choice among alternative actions with which to meet those contingencies. Even the most controlled mathematical behavior has some internal regulation associated with it.

Self-determination theory suggests that three primary human needs govern this interaction of the social and individual motivations: (a) the need for competence, (b) autonomy, and (c) relatedness. Note that both competence and relatedness are social in nature, defining what behaviors and outcomes denote success in mathematics, who is successful, and how groups can support each other (or not) to improve the success of individual members.

2.5.5 Key Individual Motivational Factors

Interest and preferences. Interest has been found to be one of the most significant predictors of mathematical achievement and persistence over the years. Interest, like most motivation factors, has both a short-term and a long-term manifestation. In the short term, the self-regulatory process shown in Fig. 2 helps the learner create an initial situational interest (Ainley and Hidi 2014), optimizing the challenge and control affordances of the task (Middleton and Toluk 1999). Over time, tasks that have consistently been evaluated as interesting situationally can be consolidated into a long-term dispositional structure: personal interest. Situational interest predicts depth of cognitive processing of information (Linnenbrink-Garcia et al. 2013), degree of positive affect (Ely et al. 2013), and effort regulation (Lee et al. 2014). Personal interest predicts continued mathematics and mathematics-intensive course taking (Lichtenberger and George-Jackson 2013), mathematics achievement in the long term (Murayama et al. 2013), and perhaps most importantly, mathematics identity (Mangu et al. 2015)—that aspect of ourselves that integrates our self-concept and cultural role with our conception of mathematics and what makes one mathematically adept (Boaler and Greeno 2000).

Perceived instrumentality. So far we have established that people engage with mathematics because it fulfills some situational interest. But interest alone cannot explain mathematics engagement. After all, many, if not most, learners display neither situational interest nor personal interest sufficient to warrant their engagement. Other factors, therefore, must play a prominent role. Among those factors is utility, the degree to which the learner feels that engagement will result in some valuable knowledge, skill, or social standing, which may be instrumental to fulfilling some non-mathematical interest (Husman et al. 2004). Like interest, this perceived instrumentality has both a short-term and a long-term manifestation. In the short term, a person may perceive the mathematical content to be useful for learning some important domain within which the task is contextualized. Such perceptions are termed *endogenous instrumentality* because they are intrinsic to the task: completing the task will result in some successful fulfillment of a short-term task-related goal, such as learning about population growth. In the long-term, a task may not have endogenous utility, but may be important for achieving more distal goals. Such perceptions are termed *exogenous instrumentality* because the immediate task is only a stepping stone to some longer-term outcome. A prime example of this is students taking calculus merely because it is pre-requisite for medical school. The immediate content is not seen as useful, but doing it successfully is critical for achieving the longer-term life goal.

Perceptions of endogenous instrumentality have been shown to increase interest in the task and the effort students are willing to apply to mathematical inquiry and ultimately to mathematical achievement (Shell et al. 2013). Exogenous instrumentality, for its part, is associated with persistence towards distal goals such as college and career choices (Nelson et al. 2015).

Personal goals. Motivation on the individual level is goal-oriented. Therefore the types of goals, their object, and the process of resolving them drive much of our academic effort. It must be stressed that goals under this framework are the learner's goals, not necessarily the teacher's or school's. The literature on goals and their role in mathematics motivation is too voluminous to review adequately here. Suffice it to say that both task-level goals (short-term) and goal orientations (long-term) greatly impact students' motivation by affecting the desired outcomes of their imagined engagement (Sheeran et al. 2005). Personal goals can be described as the interaction of three dimensions: (a) goal proximity, (b) goal specificity, and (c) goal focus.

Goal proximity. The closer a goal is to the immediate task at hand (proximity), the more learners tend to recruit strategies for self-regulation such as time management and regulation of effort (Hester 2012; Harber et al. 2003; Horstmanshof and

Zimitat 2007; Zimbardo and Boyd 1999). These strategies all increase in intensity as a function of the student's ability to see how their current engagement contributes to a more distal goal (Zhang et al. 2011). To coordinate proximal goals towards longer-term outcomes, it is critical that learners develop plans so that they can see how the mathematics content contributes to their personal interests and long-term identity.

Goal specificity. For both proximal and distal goals, the development of a coherent plan requires making goals specific. General goals such as "doing well in life" do not provide direct information for how to behave, what strategies to recruit, and who to align with when engaged in mathematics. Goals such as "learning factoring," a proximal goal, if articulated into a larger plan leading towards "becoming a mechanical engineer" provides evidence of the exogenous utility of one's engagement, and, thus, the longer-term plan guides the learner's behavior directly. Research shows that teachers can encourage students to create such plans and articulate their goals, leading to productive engagement patterns. (Ford 1992; Latham and Locke 1991; Harackiewicz and Sansone 1991).

Self-efficacy. When students engage in tasks that provide fulfillment of their learning goals, they develop a stronger sense of self-efficacy in mathematics than those who find little success or those who tend to rely on ego/performance goals for attribution of success. Self-efficacy is just one aspect of mathematical identity and interacts significantly with the social rules and norms governing what constitutes mathematical success in the classroom (Usher 2009). As such, students may develop efficacious beliefs for very different mathematical experiences. In fact, we often find great fluctuation in mathematical self-efficacy year by year (Phan 2012; Mangu et al. 2015), indicating that mathematical experiences change according to classroom, teacher, and social variables. But overall the research shows that students who develop higher mathematical self-efficacy tend to show greater interest, effort, persistence, help-seeking behavior, and, ultimately, greater mathematics achievement than those who feel their efforts in mathematics have less efficacy (Skaalvik et al. 2015).

Affect. Any task or pursuit in which the student is engaged contains affective and, particularly, emotional content. The "control-value theory" of achievement emotions (Pekrun 2006) holds that learners' expectations that application of effort will lead to success interacts with their beliefs about the perceived value of being successful to create the kinds of anticipatory structures presented in Fig. 2. When asked to engage in mathematics, learners generate anticipatory emotions such as hope or anxiety that direct the cognitive appraisal of their engagement (Goldin 2000, 2014). Reflecting on successes or non-successes results in emotional responses such as pride (success, high-value task), boredom (success, low-value task), anger/frustration (failure, high-value task), or apathy (failure, low-value task).

The reflection phase of engagement results in encoded affective structures that serve as templates for engagement on subsequent occasions. Structures such as mathematical intimacy, mathematical integrity, and math anxiety, which introduce meta-affective contexts for mathematical engagement (DeBellis and Goldin 1999, 2006), are tied to levels of situational interest. Such structures impact students' cognitive and social processing and, more long-term, their personal interests and identities in relation to mathematics.

2.5.6 Social Factors

Insufficiently addressed in the research literature in mathematical motivation is consideration of social motivation, or socio-mathematical motivation, wherein students' reasons for engagement are tied to both cognitive/affective appraisal of the mathematics content and appraisal of their role in the culture of the learning environment (Patrick et al. 2007).

Webel (2013), for example, established that engagement behaviors are often stable at the group level, conforming to the socio-mathematical/motivational norms of the class. But at the individual level, students' personal goals related to asserting or maintaining self-worth, their differing achievement goals, and non-mathematical goals may conflict with those of the group and may have even greater value to individuals than their mathematical goals, resulting in great variation in manifested motivation and engagement practices within small groups.

Thus, at a certain level, self-regulation, self-determination, and individual differences theories have the wrong focus. It is really the interaction among social and individual goals and behaviors and the requirements for engagement in mathematics tasks and pursuits that drives the depth and focus of engagement in learners. Such interactions are being studied under the umbrella concept of socially-shared regulation of learning (Panadero and Järvelä 2015). It is in this intersection of motivation, affect, and human interaction, we think, that the future of the field lies.

3 Summary and Looking Ahead

This publication's aim was to describe the current state of the art in research on mathematics-related affect. Towards this goal, experts in the different strands of research present and discussed the following relevant concepts: attitude towards mathematics, self-efficacy beliefs, teacher beliefs, mathematical identities, and mathematical motivation.

In each section, relevant findings were highlighted. Here we discuss the open questions emerging from the sections and sketch directions for further research.

In Sect. 2.1 Di Martino briefly described the historical development of the research on attitude towards math (for a more complete report see Di Martino and Zan 2015), but this development has to continue: new issues and new goals emerge. Without claiming to be exhaustive, three main directions for future research on attitude are: (a) research aimed at better describing the different profiles of attitude towards mathematics, developing new instruments, and studying the origin of certain profiles, in particular the role of didactical, social, and cultural aspects in the development of recurrent profiles of attitude towards mathematics (i.e., comparing attitudes of students from different countries, cultures, and school systems); (b) research aimed at analyzing adults' attitudes towards mathematics, in particular attitudes of in-service and future teachers, and how they influence in-service teachers' didactical choices and future teachers' professional development;

and (c) intervention research aimed at overcoming students' or teachers' negative attitudes towards mathematics. Longitudinal studies appear to be crucial to evaluate the effects of "remedial intervention on attitude" over time.

In Sect. 2.2 Pantziara discusses in detail research findings concerning self-efficacy beliefs. Researchers in the field of student mathematics self-efficacy beliefs suggest several topics that need future investigation. Particularly, there is a need for a clearer understanding of how efficacy beliefs develop so that appropriate interventions can be developed to improve student self-efficacy and collective efficacy beliefs. Moreover, the directionality of the relationship between collective efficacy and group performance remains largely unexplored. Cross-cultural research is needed, as it may reveal how mathematics self-efficacy operates in diverse contexts and how students in contrasting settings function. We need more research on self-regulation and how to improve learners' self-regulatory skills. Moreover, we need research on students' beliefs in their capabilities to exercise control over their learning environment in order to optimize their efforts. Last, there is a need for more longitudinal studies and studies with experimental designs (Davis-Kean et al. 2008; Klassen and Krawchuk 2009; Klassen and Usher 2010; Skaalvik and Skaalvik 2011).

In Sect. 2.3 Zhang and Morselli sketched a holistic and historical review of research on teacher beliefs in mathematics education, showing that the research focus has moved from the definition and characterization of beliefs to two crucial issues: the relationship between beliefs and practice and belief change. Around these two issues many other factors including internal and external contexts are also discussed widely. The first issue may be rephrased, seeing teachers as sensible systems that act in a coherent way. This suggests the need for further research aimed at a deeper comprehension of the context within which the teaching and learning takes place and of all the factors that affect teaching. The context can include curriculum reform, social-cultural influence, and also many internal inter-related dimensions (Hannula 2012). More generally, further exploration is needed to uncover possible factors that affect practice. On the one hand, research should go on investigating the roots of observed practice, as advocated by Cross (2015). On the other hand, research could address the teaching of different mathematical topics and investigate which beliefs affect such practice. Moreover, the investigation of factors affecting practice may also serve as a basis for efficient teacher development programs. Teacher education programs should move from changing beliefs per se to making teachers aware of beliefs and other factors affecting practice. New generations grow up within e-learning environments and diverse and interactive learning material and their learning environment may be very different from the school their teachers studied in. Teacher beliefs about the integration of ICT and mathematics teaching and how teachers face the changing classroom environment are also issues that need further investigation.

In Sect. 2.4, Heyd-Metzuyanim, Lutovac, and Kaasila discuss the issue of mathematical identity. In general, they note that the theoretical frameworks drawn on in the research on student and teacher identities are quite similar. Both draw heavily on socio-cultural theories that view learning as becoming a participant in a community of practice. Regardless of how the concept has been defined, most studies break the individual versus social dichotomy, trying to portray identity as personal, but also constructed in relationships with others. Notably, though, the student identity literature more often links issues of identity with race, gender, and ethnicity while the teacher identity literature mostly concentrates on the process of becoming a professional teacher. In line with the heavy reliance on socio-cultural theories, most studies of both student and teacher identity use qualitative methods for their investigation, with a major emphasis on interviews. In both of these domains, the reliance on interviews dictates a certain detachment from the actual activity in the classroom, as well as difficulty in generalizing from small sample sizes. Heyd-Metzuyanim, Lutovac, and Kaasila identify a curious absence of studies looking at both student and teacher identity. Such a focus may be of interest, especially in places of shifting participation structures, such as those where "reform" in mathematical instruction is introduced. Though the concept of identity seems to function as a nexus of social narratives and subjective experience, most literature reviewed here on both student and teacher identity is quite detached from studies dealing with emotions (e.g., mathematics anxiety), attitudes, or beliefs. This may be a result of the different theoretical frameworks drawn upon, including different methods and tools for obtaining data. Previous work done in IGPME (Frade et al. 2010) has shown the potential of examining intersections between identity and other affect-related constructs. We thus recommend pursuing this line of study.

The contribution of Middleton, Jansen, and Goldin (Sect. 2.5) shows that the reasons learners have for "playing the game" of school mathematics differ along a number of important factors. These authors find three emergent themes. The first theme concerns the time scale of the motivation structure: In the moment versus long term (Fig. 3). Situational interest and state-level preferences such as endogenous perceived instrumentality, task-level goals, and task-based efficacy beliefs interact with local affect and the norms and group-level goals of classmates to create highly individualized learner motivations at any given time. Yet, the commonality of curriculum, norms of practice, and societal norms over time yield a smaller set of long-term engagement patterns. Individual interest, exogenous instrumentality, goal orientations, and broader academic self-efficacy currently tend to lessen learners' enjoyment of and persistence in challenging mathematics as they grow older (Frenzel et al. 2010; Mangu et al. 2015; Watt 2004; Fredricks and Eccles 2002).

Middleton, Jansen, and Goldin point out that the interaction among individual-level motivational variables is not trivial. There is considerable literature showing, for example, that the variables reviewed in their contribution are indeed separable statistically, meaning that they can be modeled as more or less independent from each other (e.g., Mitchell 1993). Yet in reality most studies of these factors show relatively little variability accounted for, either separate or in conjunction (e.g., Middleton 2013; Mangu et al. 2015). There appears to be promise in assuming that goals, utility, interests, self-efficacy, and affect interact nonlinearly, and that social systems are key supports and catalysts for productive motivational sets to both (a) be developed over time and (b) find the right setting for engagement in the right moment.

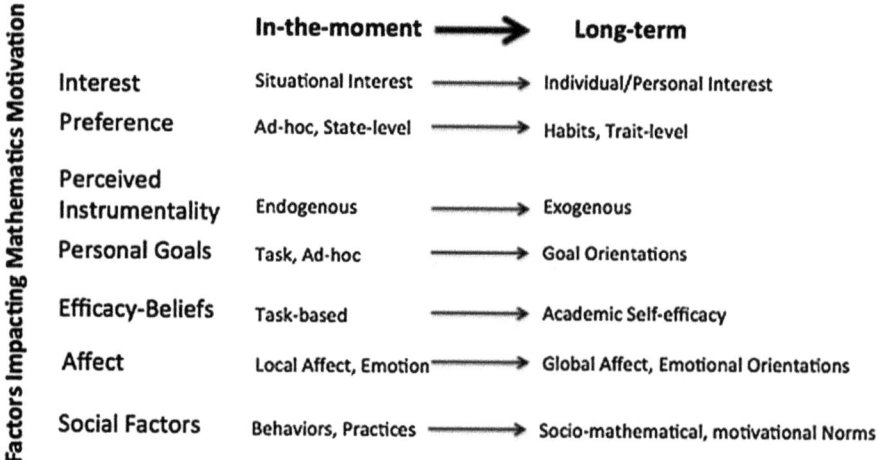

Fig. 3 Relationship of motivational variables: short term to long term

The in-the-moment perspective taken by Middleton, Jansen, and Goldin should become more of a focus for motivation research. Engagement in the moment is a place where educators may have some control over the eliciting conditions for the development of interest and goals, instrumentality and efficacy beliefs, prosocial behaviors, and productive affective structures. A retrospective look at what students have experienced in the past cannot capture the richness of interactions that have surely occurred on a day-to-day basis and that are fundamental to human beings' growth in mathematics and associated subject matter. This focus may help researchers gain traction in resolving the paradox of mathematics' unique role in promoting an informed and creative society while at the same time being one the least favorite subjects of many students.

References

Ainley, M. (2006). Connecting with learning: Motivation, affect and cognition in interest processes. *Educational Psychology Review, 18*(4), 391–405.

Ainley, M., & Hidi, S. (2014). Interest and enjoyment. In R. Pekrun & L. Linnenbrink-Garcia (Eds.), *International handbook of emotions in education* (pp. 205–227). New York: Routledge.

Ainley, M., Hidi, S., & Berndorff, D. (2002). Interest, learning, and the psychological processes that mediate their relationship. *Journal of Educational Psychology, 94*(3), 545.

Andersson, A., Valero, P., & Meaney, T. (2015). "I am [not always] a maths hater": Shifting students' identity narratives in context. *Educational Studies in Mathematics, 90*(2), 143–161.

Andrews, P., & Xenofontos, C. (2015). Analysing the relationship between the problem-solving related beliefs, competence and teaching of three Cypriot primary teachers. *Journal for Mathematics Teacher Education, 18*, 299–325.

Bandura, A. (1997). *Self-Efficacy: The exercise of control*. NY: W. H. Freeman and Company.

Bandura, A. (1986). *Social foundations of thought and action: A social cognitive theory*. Englewood Cliffs, NJ: Prentice Hall.

Beswick, K. (2012). Teachers' beliefs about school mathematics and mathematicians' mathematics and their relationship to practice. *Educational Studies in Mathematics, 79*(1), 127–147.

Beswick, K., & Callingham, R. A. (2014). The beliefs of pre-service primary and secondary mathematics teachers, in-service mathematics teachers, and mathematics teacher educators. In P. Liljedahl, C. Nicol, S. Oesterle & D. Allan (Eds.), *Proceedings of the 38th conference of the IGPME and the 36th conference of the PME-NA* (Vol. 1, pp. 137–144). Vancouver, Canada: PME.

Bishop, J. P. (2012). She's always been the smart one. I've always been the dumb one: Identities in the mathematics classroom. *Journal for Research in Mathematics Education, 43*(1), 34–74.

Black, L., Mendick, H., & Solomon, Y. (Eds.). (2009a). Mathematical relationships in education: Identities and participation. New York: Routledge.

Black, L., Williams, J., Hernandez-Martinez, P., Davis, P., Pampaka, M., & Wake, G. (2009b). Developing a "leading identity": The relationship between students' mathematical identities and their career and higher education aspirations. Educational Studies in Mathematics, 73(1), 55–72.

Boaler, J., & Greeno, J. (2000). Identity, agency, and knowing in mathematics worlds. In J. Boaler (Ed.), *Multiple perspectives on mathematics education* (pp. 171–200). Westport, CT: Ablex.

Boekaerts, M., Pintrich, P. R., & Zeidner, M. (Eds.). (2005). *Handbook of self-regulation*. Amsterdam, NL: Elsevier.

Brown, T., & McNamara, O. (2011). *Becoming a mathematics teacher: Identity and identifications*. Dordrecht, The Netherlands: Springer.

Charalambous, C. Y., Panaoura, A., & Philippou, G. (2009). Using the history of mathematics to induce changes in preservice teachers' beliefs and attitudes: Insights from evaluating a teacher education program. *Educational Studies in Mathematics, 71*, 161–180.

Chval, K. B., Arbaugh, F., Lannin, J. K., van Garderen, D., Cummings, L., et al. (2010). The transition from experienced teacher to mathematics coach: Establishing a new identity. *Elementary School Journal, 111*(1), 191–216.

Cobb, P., Gresalfi, M., & Hodge, L. L. (2009). An interpretive scheme for analyzing the identities that students develop in mathematics classrooms. *Journal for Research in Mathematics Education, 40*(1), 40–68.

Cross, D. I. (2009). Alignment, cohesion, and change: Examining mathematics teachers' belief structures and their influence on instructional practices. *Journal of Mathematics Teacher Education, 12*, 325–346.

Cross, D. I. (2015). Dispelling the notion of inconsistencies in teachers' mathematics beliefs and practices: A 3-year case study. *Journal of Mathematics Teacher Education, 18*, 173–201.

Daskalogianni, K., & Simpson, A. (2000). Towards a definition of attitude: The relationship between the affective and the cognitive in pre-university student. In T. Nakahara & M. Koyama (Eds.), *Proceedings of the 24th conference of the IGPME* (Vol. 2, pp. 217–224). Hiroshima, Japan: PME.

Davis-Kean, P., Huesmann, R., Jager, J., Collins, A., Bates, J., & Lansford, J. (2008). Changes in the relation of self-efficacy beliefs and behaviors across development. *Child Development, 79*(5), 1257–1269.

da Ponte, J. P., Oliveira, H., & Varandas, J. M. (2002). Development of pre-service mathematics teachers' professional knowledge and identity in working with information and communication technology. *Journal of Mathematics Teacher Education, 5*(2), 93–115.

DeBellis, V. A., & Goldin, G. A. (1999). Aspects of affect: Mathematical intimacy, mathematical integrity. In O. Zaslavsky (Ed.), *Proceedings of the 23rd conference of the IGPME, Haifa, Israel*, (Vol. 2, pp. 249–256). Haifa, Israel: PME.

DeBellis, V. A., & Goldin, G. A. (2006). Affect and meta-affect in mathematical problem solving: A representational perspective. *Educational Studies in Mathematics, 63*, 131–147.

Deci, E. L., & Ryan, R. M. (2008). Facilitating optimal motivation and psychological well-being across life's domains. *Canadian Psychology/Psychologie canadienne, 49*(1), 14.

de Freitas, E. (2008). Troubling teacher identity: Preparing mathematics teachers to teach for diversity. *Teaching Education, 19*(1), 43–55.

Di Martino, P., & Zan, R. (2001). Attitude toward mathematics: Some theoretical issues. In M. van den Heuvel-Panhuizen (Ed.), *Proceedings of the 25th conference of the IGPME* (Vol. 3, pp. 351–358). Utrecht, The Netherlands.

Di Martino, P., & Zan, R. (2010). 'Me and maths': Towards a definition of attitude grounded on students' narratives. *Journal of Mathematics Teacher Education, 13*(1), 27–48.

Di Martino, P., & Zan, R. (2011). Attitude towards mathematics: A bridge between beliefs and emotions, *ZDM—The International Journal on Mathematics Education, 43*(4), 471–482.

Di Martino, P., & Zan, R. (2015). The construct of attitude in mathematics education. In B. Pepin & B. Roesken-Winter (Eds.), *From beliefs to dynamic affect systems in mathematics education. Exploring a mosaic of relationships and interactions* (pp. 51–72). New York: Springer.

Drake, C., Spillane, J. P., & Hufferd-Ackles, K. (2001). Storied identities: Teacher learning and subject-matter context. *Journal of Curriculum Studies, 33*(1), 1–23.

Dutton, W. (1951). Attitudes of prospective teachers toward arithmetic. *The Elementary School Journal, 42*, 84–90.

Else-Quest, N. M., Hyde, J. S., & Linn, M. C. (2010). Cross-national patterns of gender differences in mathematics: A meta-analysis. *Psychological Bulletin, 136*(1), 103.

Ely, R., Ainley, M., & Pearce, J. (2013). More than enjoyment: Identifying the positive affect component of interest that supports student engagement and achievement. *Middle Grades Research Journal, 8*(1), 13.

Erens, R., & Eichler, A. (2014). On the structure of secondary high-school teachers' belief systems on calculus. In P. Liljedahl, C. Nicol, S. Oesterle & D. Allan (Eds.), *Proceedings of the 38th conference of the IGPME and the 36th conference of the PME-NA* (Vol. 3, pp. 25–32). Vancouver, Canada: PME.

Fang, Z. (1996). A review of research on teacher beliefs and practices. *Educational Research, 38*(1), 47–65.

Fennema, E. (1989). The study of affect and mathematics: A proposed generic model for research. In D. McLeod & V. Adams (Eds.), *Affect and mathematical problem solving. A new perspective* (pp. 205–219). New York: Springer-Verlag.

Fenstermacher, G. (1979). A philosophical consideration of recent research on teacher effectiveness. In L. Shulman (Ed.), *Review of research in education* (pp. 157–185). Itasca, IL: Peacock.

Ford, M. E. (1992). *Motivating humans: Goals, emotions, and personal agency beliefs.* Sage Publications.

Frade, C., Roesken, B., & Hannula, M. S. (2010). Identity and affect in the context of teachers' professional development. In M. M. F. Pinto & T. F. Kawasaki (Eds.), *Proceedings of the 34th conference of the IGPME* (Vol. 1, pp. 247–279). Belo Horizonte, Brazil: PME.

Fredricks, J. A., & Eccles, J. S. (2002). Children's competence and value beliefs from childhood through adolescence: Growth trajectories in two male-sex-typed domains. *Developmental psychology, 38*(4), 519–533.

Frenzel, A. C., Goetz, T., Pekrun, R., & Watt, H. M. (2010). Development of mathematics interest in adolescence: Influences of gender, family, and school context. *Journal of Research on Adolescence, 20*(2), 507–537.

Frenzel, A. C., Pekrun, R., & Goetz, T. (2007). Perceived learning environment and students' emotional experiences: A multilevel analysis of mathematics classrooms. *Learning and Instruction, 17*, 478–493.

Furinghetti, F. (1997). On teachers' conceptions: From a theoretical framework to school practice. In G. A. Makrides (Ed.), *Proceedings of the first Mediterranean conference on mathematics* (pp. 277–287). Cyprus: Cyprus Pedagogical Institute and Cyprus Mathematical Society.

Furinghetti, F., & Morselli, F. (2011). Beliefs and beyond: Hows and whys in the teaching of proof. *ZDM—The International Journal of Mathematics Education, 43*, 587–599.

Furinghetti, F., & Pehkonen, E. (2002). Rethinking characterizations of beliefs. In G. C. Leder, E. Pehkonen, & G. Törner (Eds.), *Beliefs: A hidden variable in mathematics education?* (pp. 39–57). Dordrecht: Kluwer.

Gee, J. P. (2001). Identity as an analytic lens for research in education. *Review of Research in Education, 25*, 99–125.

Gellert, U., Espinoza, L., & Barbé, J. (2013). Being a mathematics teacher in times of reform. *ZDM, 45*(4), 535–545.

Germann, P. (1988). Development of the attitude toward science in school assessment and its use to investigate the relationship between science achievement and attitude toward science in school. *Journal of Research in Science Teaching, 25*(8), 689–703.

Glaser, B., & Strauss, A. (1967). *The discovery of grounded theory. Strategies for qualitative research*. Chicago: Aldine.

Goldin, G. A. (2000). Affective pathways and representation in mathematical problem solving. *Mathematical thinking and learning, 2*(3), 209–219.

Goldin, G. A. (2014a). Perspectives on emotion in mathematical engagement, learning, and problem solving. In R. Pekrun & L. Linnenbrink-Garcia (Eds.), *Handbook of emotions in education* (pp. 391–414). New York, NY: Taylor & Francis.

Goldin, G. A., Epstein, Y. M., Schorr, R. Y., & Warner, L. B. (2011). Beliefs and engagement structures: Behind the affective dimension of mathematical learning. *ZDM, 43*, 547–556.

Goos, M. (2005). A sociocultural analysis of the development of pre-service and beginning teachers' pedagogical identities as users of technology. *Journal of Mathematics Teacher Education, 8*(1), 35–59.

Goos, M. E., & Bennison, A. (2008). Developing a communal identity as beginning teachers of mathematics: Emergence of an online community of practice. *Journal of Mathematics Teacher Education, 11*(1), 41–60.

Graven, M. (2005). Mathematics teacher retention and the role of Identity: Sam's story. *Pythagoras, 61*, 2–10.

Green, T. F. (1971). *The activities of teaching*. New York: McGraw-Hill.

Grootenboer, P. (2008). Mathematical belief change in prospective primary teachers. *Journal of Mathematics Teacher Education, 11*, 479–497.

Grootenboer, P. (2013). The praxis of mathematics teaching: Developing mathematical identities. *Pedagogy, Culture and Society, 21*(2), 321–342.

Guberman, R., & Leikin, R. (2013). Interest and difficulty: Changes in teachers' views of multiple solution tasks. *Journal of Mathematics Teacher Education, 16*(1), 33–56.

Hannula, M. S. (2002). Attitude toward mathematics: Emotions, expectations and values. *Educational Studies in Mathematics, 49*, 25–46.

Hannula, M. S. (2011). The structure and dynamics of affect in mathematical thinking and learning. In M. Pytlak, T. Rowland & E. Swoboda (Eds.), *Proceedings of the 7th conference of ERME* (pp. 34–60). Rzeszów, Poland.

Hannula, M. S. (2012). Exploring new dimensions of mathematics-related affect: Embodied and social theories. *Research in Mathematics Education, 14*(2), 137–161.

Hannula, M. S. (2015). Emotions in problem solving. In S. J. Cho (Ed.), (2015). *Selected regular lectures from the 12th international congress on mathematical education* (pp. 269–288). Springer.

Hannula, M. S., Bofah, E., Tuohilampi, L., & Metsämuuronen, J. (2014). A longitudinal analysis of the relationship between mathematics-related affect and achievement in Finland. In S. Oesterle, P. Liljedahl, C. Nicol & D. Allan (Eds.), *Proceedings of the 38th conference of the IGPME and the 36th conference of the PME-NA* (Vol. 3, pp. 249–256). Vancouver, Canada: PME.

Harackiewicz, J. M., & Sansone, C. (1991). Goals and intrinsic motivation: You can get there from here. *Advances in Motivation and Achievement, 7*, 21–49.

Harber, K. D., Zimbardo, P. G., & Boyd, J. N. (2003). Participant self-selection biases as a function of individual differences in time perspective. *Basic and Applied Social Psychology, 25*(3), 255–264.

Harmon-Jones, E., Harmon-Jones, C., & Price, T. F. (2013). What is approach motivation? *Emotion Review, 5*(3), 291–295.

Harré, R., & van Langenhove, L. (1999). *Positioning theory*. Oxford: Blackwell.

Hart, L. (1989). Describing the affective domain: Saying what we mean. In D. McLeod & V. Adams (Eds.), *Affect and mathematical problem solving* (pp. 37–45). New York: Springer.

Hester, A. (2012). The effect of personal goals on student motivation and achievement. *Studies in teaching: 2012 research digest. Action research projects presented at annual research forum*. Winston-Salem, North Carolina.

Heyd-Metzuyanim, E. (2015). Vicious cycles of identifying and mathematizing—a case study of the development of mathematical failure. *Journal of the Learning Sciences, 24*(4), 504–549.

Heyd-Metzuyanim, E., & Sfard, A. (2012). Identity struggles in the mathematics classroom: On learning mathematics as an interplay of mathematizing and identifying. *International Journal of Educational Research, 51–52*, 128–145.

Hodgen, J., & Askew, M. (2007). Emotion, identity and teacher learning: Becoming a primary mathematics teacher. *Oxford Review of Education, 33*(4), 469–487.

Hodges, T. E., & Cady, J. A. (2012). Negotiating contexts to construct an identity as a mathematics teacher. *Journal of Educational Research, 105*(2), 112–122.

Holland, D., Lachicotte, W., Skinner, D., & Cain, C. (1998). *Identity and agency in cultural worlds*. Harvard University Press.

Horstmanshof, L., & Zimitat, C. (2007). Future time orientation predicts academic engagement among first-year university students. *British Journal of Educational Psychology, 77*(3), 703–718.

Hossain, S., Mendick, H., & Adler, J. (2013). Troubling "understanding mathematics in-depth": Its role in the identity work of student-teachers in England. *Educational Studies in Mathematics, 84*(1), 35–48.

Hulleman, C. S., Godes, O., Hendricks, B. L., & Harackiewicz, J. M. (2010). Enhancing interest and performance with a utility value intervention. *Journal of Educational Psychology, 102*(4), 880.

Husman, J., & Hilpert, J. (2007). The intersection of students' perceptions of instrumentality, self-efficacy, and goal orientations in an online mathematics course. *Zeitschrift für Pädagogische Psychologie, 21*(3/4), 229–239.

Husman, J., Derryberry, W. P., Crowson, H. M., & Lomax, R. (2004). Instrumentality, task value, and intrinsic motivation: Making sense of their independent interdependence. *Contemporary Educational Psychology, 29*(1), 63–76.

Jiang, Y., Song, J., Lee, M., & Bong, M. (2014). Self-efficacy and achievement goals as motivational links between perceived contexts and achievement. *Educational Psychology, 34*(1), 92–117.

Jones, L., Brown, T., Hanley, U., & McNamara, O. (2000). An enquiry into transitions: From being a 'learner of mathematics' to becoming a 'teacher of mathematics'. *Research in Education, 63*, 1–10.

Kaasila, R. (2007). Using narrative inquiry for investigating the becoming of a mathematics teacher. *ZDM—The International Journal on Mathematics Education, 39*(3), 205–213.

Kaasila, R., Hannula, M. S., Laine, A., & Pehkonen, E. (2008). Socio-emotional orientations and teacher change. *Educational Studies in Mathematics, 67*(2), 111–123.

Karsenty, R., & Vinner, S. (2000). What do we remember when it's over? Adults' recollections of their mathematical experience. In T. Nakahara & M. Koyama (Eds.), *Proceedings of the 24th conference of the IGPME* (Vol. 4, pp. 451–458). Hiroshima, Japan: PME.

Klassen, R., & Krawchuk, L. (2009). Collective motivation beliefs of early adolescents working in small groups. *Journal of School Psychology, 47*, 101–120.

Klassen, R., & Usher, E. (2010). Self-efficacy in educational settings: Resent research and emerging directions. In T. Urdan & S. Karabenick (Eds.), *The decade ahead: Theoretical perspectives on motivation and achievement* (pp. 1–34). Bingley, UK: Emerald Group Publishing Limited.

Latham, G. P., & Locke, E. A. (1991). Self-regulation through goal setting. *Organizational Behavior and Human Decision Processes, 50*(2), 212–247. doi:10.1016/0749-5978(91)90021-K.

Lave, J., & Wenger, E. (1991). *Situated learning: Legitimate peripheral participation*. New York: Cambridge University Press.

Leatham, K. (2006). Viewing mathematics teachers' beliefs as sensible systems. *Journal of Mathematics Teacher Education, 9*, 91–102.

Leder, G. (1985). Measurement of attitude to mathematics. *For the Learning of Mathematics, 5*(3), 18–22.

Lee, W., Lee, M. J., & Bong, M. (2014). Testing interest and self-efficacy as predictors of academic self-regulation and achievement. *Contemporary Educational Psychology, 39*(2), 86–99.

Lerman, S. (2000). The social turn in mathematics education research. In J. Boaler (Ed.), *Multiple perspectives on mathematics teaching and learning* (pp. 19–44). Westport, CN: Ablex.

Leung, F. K. (2001). In search of an East Asian Identity in Mathematics Education. *Educational Studies in Mathematics, 47*(1), 35–51.

Lichtenberger, E., & George-Jackson, C. (2013). Predicting high school students' interest in majoring in a STEM field: Insight into high school students' postsecondary plans. *Journal of Career and Technical Education, 28*(1), [article available online: https://ejournals.lib.vt.edu/index.php/JCTE/article/view/567/598].

Liljedahl, P. (2010). Noticing rapid and profound mathematics teacher change. *Journal of Mathematics Teacher Education, 13*(5), 411–423.

Liljedahl, P. (2011). The theory of conceptual change as a theory for changing conceptions. *Nordic Studies in Mathematics Education, 16*(1–2), 101–124.

Liljedahl, P., Oesterle, S., & Bernèche, C. (2012). Stability of beliefs in mathematics education: A critical analysis. *Nordic Studies in Mathematics Education, 17*(3–4), 101–118.

Lim, J. H. (2008). Adolescent girls' construction of moral discourses and appropriation of primary identity in a mathematics classroom. *ZDM, 40*(4), 617–631.

Linnenbrink-Garcia, L., Patall, E. A., & Messersmith, E. E. (2013). Antecedents and consequences of situational interest. *British Journal of Educational Psychology, 83*(4), 591–614.

Lutovac, S., & Kaasila, R. (2011). Beginning a pre-service teacher's mathematical identity work through narrative rehabilitation and bibliotherapy. *Teaching in Higher Education, 16*(2), 225–236.

Lutovac, S., & Kaasila, R. (2014). Pre-service teachers' future-oriented mathematical identity work. *Educational Studies in Mathematics, 85*(1), 129–142.

Ma, J. Y., & Singer-Gabella, M. (2011). Learning to teach in the figured world of reform mathematics: Negotiating new models of identity. *Journal of Teacher Education, 62*(1), 8–22.

Mangu, D., Lee, A., Middleton, J. A., & Nelson, J. K. (2015). Motivational Factors Predicting STEM and Engineering Career Intentions for High School Students. *2015 IEEE frontiers in education conference proceedings* (pp. 2285–2291). IEEE: El Paso, TX.

Martin, D. B. (2007). Mathematics learning and participation in African American context: The co-construction of identity in two intersecting realms of experience. In N. S. Nasir & P. Cobb (Eds.), *Improving access to mathematics*. New York and London: Teachers College Press.

McLeod, D. (1992). Research on affect in mathematics education: A reconceptualization. In D. A. Grouws (Ed.), *Handbook of research on mathematics teaching and learning* (pp. 575–596). New York, NY: Macmillan.

McLeod, D., & Adams, V. (Eds.). (1989). *Affect and mathematical problem solving. A new perspective*. New York: Springer-Verlag.

Middleton, J. A. (1995). A study of intrinsic motivation in the mathematics classroom: A personal constructs approach. *Journal for Research in Mathematics Education, 26*(3), 254–279.

Middleton, J. A. (2013). More than motivation: The combined effects of critical motivational variables on middle school mathematics achievement. *Middle Grades Research Journal, 8*(1), 77–95.

Middleton, J. A., Jansen, A., & Goldin, G. E. (in press). The complexities of mathematical engagement: Motivation, affect, and social interactions. In J. Cai (Ed.), *As yet untitled*. Reston, VA: National Council of Teachers of Mathematics.

Middleton, J. A., Lesh, R., & Heger, M. (2002). Interest, identity, and social functioning: Central features of modeling activity. In H. Doerr & R. Lesh (Eds.), *Beyond constructivism: A models and modeling perspective on mathematics problem solving, learning and teaching* (pp. 405–431). Mahwah, NJ: Lawrence Erlbaum Associates Inc.

Middleton, J. A., & Toluk, Z. (1999). First steps in the development of an adaptive, decision-making theory of motivation. *Educational Psychologist, 34*(2), 99–112.

Mitchell, M. (1993). Situational interest: Its multifaceted structure in the secondary school mathematics classroom. *Journal of Educational Psychology, 85*(3), 424.

Moser, J. S., Moran, T. P., Schroder, H. S., Donnellan, M. B., & Yeung, N. (2013). On the relationship between anxiety and error monitoring: A meta-analysis and conceptual framework. *Frontiers in human neuroscience, 7.*

Multon, K. D., Brown, S. D., & Lent, R. W. (1991). Relation of self-efficacy beliefs to academic outcomes: A meta-analytic investigation. *Journal of Counseling Psychology, 38*(1), 30.

Murayama, K., Pekrun, R., Lichtenfeld, S., & Vom Hofe, R. (2013). Predicting long-term growth in students' mathematics achievement: The unique contributions of motivation and cognitive strategies. *Child Development, 84*(4), 1475–1490.

Nasir, N. S. (2002). Identity, goals, and learning: Mathematics in cultural practice. *Mathematical Thinking and Learning, 4*(2), 213–247.

National Science Board. (2014). *Science and engineering indicators, 2014*. Arlington, VA, USA: National Science Foundation.

Nelson, K. G., Shell, D. F., Husman, J., Fishman, E. J., & Soh, L. K. (2015). Motivational and self-regulated learning profiles of students taking a foundational engineering course. *Journal of Engineering Education, 104*(1), 74–100.

Neumayer-Depiper, J. (2013). Teacher identity work in mathematics teacher education. *For the Learning of Mathematics, 33*(1), 9–15.

Oppland-Cordell, S., & Martin, D. B. (2014). Identity, power, and shifting participation in a mathematics workshop: Latin@ students' negotiation of self and success. *Mathematics Education Research Journal, 27*(1), 21–49. doi:10.1007/s13394-014-0127-6.

Osborne, J., Simon, S., & Collins, S. (2003). Attitudes towards science: A review of the literature and its implications. *International Journal of Science Education, 25*(9), 1049–1079.

Özdemir, E., & Pape, S. (2013). The role of interactions between student and classroom context in developing adaptive self-efficacy in one sixth-grade mathematics classroom. *School Science and Mathematics, 10*, 248–258.

Pajares, F. (1996). Self-Efficacy in Academic Settings. *Review of Educational Research., 66*, 543–578.

Pampaka, M., Kleanthous, I., Hutcheson, G., & Wake, G. (2011). Measuring mathematics self-efficacy as a learning outcome. *Research in Mathematics Education, 13*, 169–190.

Panadero, E., & Järvelä, S. (2015). Socially shared regulation of learning: A review. *European Psychologist, 20*(3), 190–203.

Patrick, H., Ryan, A. M., & Kaplan, A. (2007). Early adolescents' perceptions of the classroom social environment, motivational beliefs, and engagement. *Journal of Educational Psychology, 99*(1), 83.

Pekrun, R. (2006). The control-value theory of achievement emotions: Assumptions, corollaries, and implications for educational research and practice. *Educational Psychology Review, 18*, 315–341.

Phan, H. P. (2012). The development of English and mathematics self-efficacy: A latent growth curve analysis. *The Journal of Educational Research, 105*(3), 196–209.

Philipp, R. A. (2007). Mathematics teachers' beliefs and affect. In F. K. Lester (Ed.), *Second handbook of research on mathematics teaching and learning* (pp. 257–315). Charlotte: Information Age Publishing.

Richardson, V. (1996). The role of attitudes and beliefs in learning to teach. In J. Sikula (Ed.), *Handbook of research on teacher education* (2nd ed., pp. 102–119). New York: Macmillan.

Rokeach, M. (1968). *Beliefs, attitudes and values: A theory of organization and change.* San Francisco: Jossey-Bass.

Sadler, P. M., Sonnert, G., Hazari, Z., & Tai, R. (2012). Stability and volatility of STEM career interest in high school: A gender study. *Science Education, 96*(3), 411–427.

Sansone, C., & Thoman, D. B. (2005). Interest as the missing motivator in self-regulation. *European Psychologist, 10*(3), 175–186.

Schoenfeld, A. H. (2011). Toward professional development for teachers grounded in a theory of decision making. *ZDM—The International Journal of Mathematics Education, 43*, 457–469.

Schukajlow, S., Leiss, D., Pekrun, R., Blum, W., Müller, M., & Messner, R. (2012). Teaching methods for modelling problems and students' task-specific enjoyment, value, interest and self-efficacy expectations. *Educational Studies in Mathematics, 79*(2), 215–237.

Schunk, D. H., & Zimmerman, B. J. (Eds.). (1998). *Self-regulated learning: From teaching to self-reflective practice.* Guilford Press.

Sfard, A. (2008). *Thinking as communicating.* New York: Cambridge University Press.

Sfard, A., & Prusak, A. (2005). Telling identities: In search of an analytic tool for investigating learning as a culturally shaped activity. *Educational Researcher, 34*(4), 14–22.

Sheeran, P., Webb, T. L., & Gollwitzer, P. M. (2005). The interplay between goal intentions and implementation intentions. *Personality and Social Psychology Bulletin, 31*(1), 87–98.

Shell, D. F., Hazley, M. P., Soh, L. K., Ingraham, E., & Ramsay, S. (2013). Associations of students' creativity, motivation, and self-regulation with learning and achievement in college computer science courses. In *Frontiers in education conference, 2013 IEEE* (pp. 1637–1643). IEEE.

Simpkins, S. D., Davis-Kean, P. E., & Eccles, J. S. (2006). Math and science motivation: A longitudinal examination of the links between choices and beliefs. *Developmental Psychology, 42*(1), 70.

Skaalvik, E. M., & Skaalvik, S. (2011). Self-concept and self-efficacy in mathematics: Relation with mathematics motivation and achievement. *Journal of Education Research, 5*(3/4), 241–265.

Skaalvik, E. M., Federici, R. A., & Klassen, R. M. (2015). Mathematics achievement and self-efficacy: Relations with motivation for mathematics. *International Journal of Educational Research, 72*, 129–136.

Skott, J. (2009). Contextualising the notion of 'belief enactment'. *Journal of Mathematics Teacher Education, 12*, 27–46.

Solomon, Y. (2009a). *Mathematical literacy: Developing identities of inclusion.* London: Routledge.

Speer, N. (2005). Issues of methods and theory in the study of mathematics teachers' professed and attributed beliefs. *Educational Studies in Mathematics, 58*, 361–391.

Spillane, J. P. (2000). A fifth-grade teacher's reconstruction of mathematics and literacy teaching: Exploring interactions among identity, learning, and subject-matter. *Elementary School Journal, 100*(4), 307–330.

Swan, M. (2007). The impact of task-based professional development on teachers' practices and beliefs: A design research study. *Journal of Mathematics Teachers Education, 10*, 217–237.

Thompson, A. (1992). Teacher's beliefs and conceptions: A synthesis of the research. In D. A. Grouws (Ed.), *Handbook of research on mathematics teaching and learning* (pp. 127–146). New York: MacMillan.

Tytler, R., & Osborne, J. (2012). Student attitudes and aspirations towards science. In B. Fraser, K. Tobin, & C. McRobbie (Eds.), *Second international handbook of science education* (pp. 597–625). Netherlands: Springer.

Usher, E. L. (2009). Sources of middle school students' self-efficacy in mathematics: A qualitative investigation. *American Educational Research Journal, 46*(1), 275–314.

Usher, E. L., & Pajares, F. (2008). Sources of self-efficacy in school: Critical review of the literature and future directions. *Review of Educational Research, 78*(4), 751–796.

Usher, E., & Pajares, F. (2009). Sources of self-efficacy in mathematics: A validation study. *Contemporary Educational Psychology, 34*, 89–101.

van Zoest, L. R., & Bohl, J. V. (2005). Mathematics teacher identity: A framework for understanding secondary school mathematics teachers' learning through practice. *Teacher Development, 9*(3), 315–345.

Varelas, M., Martin, D. B., & Kane, J. M. (2013). Content learning and identity construction: A framework to strengthen African American Students' mathematics and science learning in urban elementary schools. *Human Development, 55*(5–6), 319–339. doi:10.1159/000345324.

Walshaw, M. (2004). Pre-service mathematics teaching in the context of schools: An exploration into the constitution of identity. *Journal of Mathematics Teacher Education, 7*(1), 63–86.

Watt, H. M. (2004). Development of Adolescents' Self-Perceptions, Values, and Task Perceptions According to Gender and Domain in 7th-through 11th-Grade Australian Students. *Child development, 75*(5), 1556–1574.

Webel, C. (2013). High school students' goals for working together in mathematics class: Mediating the practical rationality of studenting. *Mathematical Thinking and Learning, 15*(1), 24–57.

Wenger, E. (1998). *Communities of practice.* Cambridge: Cambridge University Press.

Wilkins, J. L. M. (2008). The relationship among elementary teachers' content knowledge, attitudes, beliefs, and practices. *Journal of Mathematics Teacher Education, 11*(2), 139–164.

Williams, T., & Williams, K. (2010). Self-efficacy and performance in mathematics: Reciprocal determinism in 33 nations. *Journal of Educational Psychology, 102*(2), 453–466.

Wong, N. Y., Ding, R., & Zhang, Q. P. (2016). From classroom environment to conception of mathematics. In R. B. King & A. B. I. Bernardo (Eds.), *The psychology of Asian learners* (pp. 541–557). Singapore: Springer.

Wood, M. B., & Kalinec, C. A. (2012). Student talk and opportunities for mathematical learning in small group interactions. *International Journal of Educational Research, 51–52*, 109–127.

Young, C. B., Wu, S. S., & Menon, V. (2012). The neurodevelopmental basis of math anxiety. *Psychological Science,* 0956797611429134.

Zan, R., Brown, L., Evans, J., & Hannula, M. (2006a). Affect in mathematics education: An introduction. *Educational Studies in Mathematics, 63*(2), 113–121.

Zhang, L., Karabenick, S. A., Maruno, S. I., & Lauermann, F. (2011). Academic delay of gratification and children's study time allocation as a function of proximity to consequential academic goals. *Learning and Instruction, 21*(1), 77–94.

Zimbardo, P. G., & Boyd, J. N. (1999). Putting time in perspective: A valid, reliable individual-differences metric. *Journal of Personality and Social Psychology, 77*(6), 1271.

Zimmerman, B. (2000). Self-efficacy: An essential motive to learn. *Contemporary Educational Psychology, 25,* 82–91.

Zimmerman, B. (2005). Attaining self-regulation: A social cognitive perspective. In M. Boekaerts, P. R. Pintrich, & M. Zeidner (Eds.), *Handbook of self-regulation* (pp. 13–41). Amsterdam, NL: Elsevier.

Further Reading on Mathematics Related Affect

Black, L., Mendick, H., & Solomon, Y. (Eds.). (2009c). *Mathematical relationships in education: Identities and participation.* New York: Routledge.

Evans, J. (2002). *Adults' mathematical thinking and emotions: A study of numerate practice* (Vol. 16). Routledge.

Evans, J., Morgan, C., & Tsatsaroni, A. (2006). Discursive positioning and emotion in school mathematics practices. *Educational Studies in Mathematics, 63*(2), 209–226.

Fives, H., & Gregoire, G. M. (Eds.). (2015). *Handbook of research on teachers' beliefs.* New York: Routledge.

Goldin, G. A. (2014b). Perspectives on emotion in mathematical engagement, learning, and problem solving. In R. Pekrun & L. Linnenbrink-Garcia (Eds.), *Handbook of emotions in education* (pp. 391–414). New York, NY: Taylor & Francis.

Leder, G. C., Pehkonen, E., & Törner, G. (Eds.). (2002). *Beliefs: A hidden variable in mathematics education?.* Dordrecht: Kluwer Academic Publishers.

Maass, J., & Schlöglmann, W. (Eds.). (2009). *Beliefs and attitudes in mathematics education: New research results.* Rotterdam: Sense.

Martin, D. B. (Ed.). (2010). *Mathematics teaching, learning, and liberation in the lives of black children.* Routledge.

Middleton, J. A., & Jansen, A. (2011). *Motivation matters and interest counts: Fostering engagement in mathematics.* Reston, VA: National Council of Teachers of Mathematics.

Pekrun, R., & Linnenbrink-Garcia, L. (Eds.). *Handbook of emotions in education* (pp. 391–414). New York, NY: Taylor & Francis.

Pepin, B., & Roesken-Winter, B. (Eds.). (2015). *From beliefs to dynamic affect systems in mathematics education.* Cham: Springer.

Skott, J., Van Zoest, L., & Gellert, U. (Eds.). (2013). Theoretical frameworks in research on and with mathematics teachers. *ZDM—The International Journal on Mathematics Education, 45*(4).

Solomon, Y. (2009b). *Mathematical literacy: Developing identities of inclusion.* London: Routledge.

Urdan & Karabenick, S. (Eds.). *The decade ahead: Theoretical perspectives on motivation and achievement* (pp. 1–34). Bingley, UK: Emerald Group Publishing Limited.

Zan, R. (2013). Solid findings on student's attitudes to mathematics. *European Mathematical Society Newsletter, 89,* 51–53.

Zan, R., Brown, L., Evans, J., & Hannula, M. (Eds.). (2006). *Educational Studies in Mathematics, 63*(2) (Special Issue: Affect in Mathematics Education).

GPSR Compliance
The European Union's (EU) General Product Safety Regulation (GPSR) is a set
of rules that requires consumer products to be safe and our obligations to
ensure this.

If you have any concerns about our products, you can contact us on

ProductSafety@springernature.com

In case Publisher is established outside the EU, the EU authorized
representative is:

Springer Nature Customer Service Center GmbH
Europaplatz 3
69115 Heidelberg, Germany

www.ingramcontent.com/pod-product-compliance
Ingram Content Group UK Ltd.
Pitfield, Milton Keynes, MK11 3LW, UK
UKHW020217231225
466357UK00011B/189